The Womb Obsession

A Theory of Sexual Preferences and Orientation

By theorist

Claude Brickell

BRICBOOKS

New York

In memory of

Sanford Meisner

a brilliant teacher, an artistic inspiration and

a personal friend…

Table of Contents

Acknowledgements

Many teachers, colleagues and friends have influenced me in the development of this theory and I am indebted to each of them for their insightful contributions.

I have included many entries in detail by others throughout the book and have credited them whenever they appeared. The authors or entities were quoted simply because their stating of the facts was well-constructed and I did not feel I could accomplish that better, myself.

Of human behavioral field's professionals, in general, I am equally in admiration of their ongoing pursuits in research and their conclusions. What I have presented here is a theory yet to be studied fully and certainly not to be taken as fact by any of them, at this point. I welcome their examination of it, however.

Begin at the beginning…

… said the king to Alice in *Alice in Wonderland…* and so I do.

Man's preoccupation with human sexuality is as old as civilization, itself. And, with his ascension to higher intelligence, illustrated for us in the Judeo-Christian myth 'Adam and Eve,' the desire to understand the multiple layers of sexuality and the fundamental motivations behind it has fascinated him for centuries, following.

Anthropologists tell us that between 8,000 and 6,000 B.C., called the Mesolithic period, man began to spread throughout the planet as ice from the Ice Age melted. But man was still wandering about in primitive existence eating berries and plants… and yes, even small animals. As he began to improve his hunting skills, he went for larger animals. That was not easy. It took the help of fellow hunters as wolves and other predators did. He needed to coordinate this new effort with other *humans*.

Up until then, man was simply communicating as most animals do using gestures, grunts and groans. To improve his skills, further, man needed to interact with others more clearly, more precisely. So, he began creating specific sounds that meant specific things as prairie dogs and other animals did, and this became language... the spoken word. Man needed this clarity to work together, effectively, as a team in order to become efficient hunters. The tribe structure followed and this was the beginning of civilization.

We see in cave paintings, marvelous early drawings, the depiction of animals which were specifically targeted for the hunt. And, in addition, depictions of fertility and the harvest as man had also begun mastering the planting of vegetables and grain, the sustenance of this new tribe. Using his unique talent among animals, which I refer to as *objective representation*, man began enactments of those essential tasks or goals through pantomime and dance, creating rituals that were based on them. These rituals became visual stories, as well, which evolved into the early myths.

In these cave depictions, we can clearly see man's growing focus on his two most fundamental urges... hunger and procreation. That great bison he wanted so much to bring down, that special birth he or she wanted to bring about... even that successful harvest that would carry the tribe through the seasons, the entire year. Everything was focused on the urge to nourish and the urge to procreate, nature's two essentials for all living beings. So, it is little wonder man's focus on the hunt (power) and on procreation (sexuality) would dominate his artistic expression. Every cave drawing, every ritual dance, every story enacted was driven by these two basic needs. We can see that preoccupation developing further in the later myths, particular those of the Minoans, which followed two to three thousand years after the Mesolithic period.

Trade dominated these ancient civilizations, and with the growing extent and sophistication of this, man needed a more precise way of keeping track of things. He began marking symbols onto moist clay plaques indicating very specific things like the name of an item traded, the number and even the price or trade exchange for it. And out of this new marked-down accounting system evolved the art of writing. By the time the Greek civilization came around (8th to 6th centuries B.C.), writing was widespread throughout the known world. The Greek writing, based on their own particular language, was undoubtedly most advanced. As a matter of fact, it formed the basis of all European writing, today.

What the Greeks did, then, which was highly developed for the time, was to take the ancient myths and write them down (e.g., *Helen of Troy*)… at least their writings of them were the most sophisticated, by that time. And they portrayed them as dramas, too. Plays like Sophocles' *Oedipus Rex* was the story of a man who unwittingly kills his father (power) and sleeps with his mother (sexuality), then blinds himself in shame when this is revealed to him. Sigmund Freud, twenty-three hundred years later, appropriated this as the prime example of the unconscious childhood desire which he dubbed the Oedipus complex. But, before Freud, power and sexuality were simply portrayed in man's creative objective representation, mentioned earlier… in novels, dance, music, drama and art. It was Freud and his followers who first brought analysis of these basic urges -- *why* we do the things we do -- to our modern consciousness.

"In creating psychoanalysis, a clinical method for treating psychopathology using dialogue between a patient and a psychoanalyst, Freud developed his therapeutic techniques like the use of free association (in which patients report their thoughts without reservation and in whatever order they might spontaneously occur) and discovered transference (the process in which patients displace onto their analysts feelings derived from the sexual experiences and fantasies of their childhood) establishing its central role in the analytic process.

Freud's redefinition of sexuality, to include its infantile forms, led him to formulate the central tenet of his psychoanalytical theory, that of the Oedipus complex. His analysis of his own and his patients' dreams as wish-fulfillments provided him with models for the clinical analysis of symptom formation and the mechanisms of repression; and further elaboration of his theory of the unconscious as an agency disruptive of conscious states of mind. Freud postulated the existence of libido, an energy with which mental process and structures are invested and which generate erotic attachments... and a death drive, the source of repetition, hate, aggression and guilt." [Wikipedia]

A contemporary of Freud and a sometime colleague was Carl Jung. "Jung was a psychotherapist and psychiatrist who founded analytical psychology. The central concept of this applies individuation, the psychological process of integrating the opposites, including the conscious with the unconscious, while, he stressed, maintaining their relative autonomy. Jung considered individuation to be the central process of human development. He created some of the best-known psychological concepts, including the archetype, the collective unconscious, the complex and synchronicity. He saw the human psyche as, 'by nature, religious' (innate) and he made this the main focus of his explorations. Jung is one of the foremost contemporary contributors to dream analysis and symbolization." [Wikipedia]

No two psychoanalytical theorists have been more instrumental in the exploration of human emotional behavior than Freud and Jung. But what about the influence of pre- and perinatal experiences on the human psyche? While it is true Freud experimented and briefly wrote about these influences, he failed to take more than a casual interest in the subject. The same could be said of Jung with the exception, possibly, of his collective unconscious as it relates to all human psyche, in general.

Although pre- and perinatal influences on the human psyche originated with Freud, it remained an undeveloped science since his time and was only seriously considered, theorized and studied beginning with the twentieth century and onward. And the field of behavioral science remains highly controversial, today.

"Although the theoretical and psychotherapeutic approaches vary in their treatment of the topic, a common thread is the fundamental importance of pre- and perinatal experiences as they play a major role in the shaping of human personality in future psychological development. Yet, somewhat contrary to the evidence, this important assertion is not widely supported in psychology. There are widespread doubts regarding the extent to which newborn infants are even capable of forming memories, the effects of any such memories on their personality and the possibility of one recovering them from an unconscious mind which, itself, is the subject of argument in the field.

Only a minority of psychologists have had direct experience of the therapeutic modalities that explore these phenomena and many question the validity and even the existence of repressed memories present in the yet unborn. However, *experience* and *memory*, together, are not synonymous, and, while a fetal infant may not be able to recall his or her experiences, he or she still lived in those moments and possibly possessed neurological, psychological or physiological responses to them which may influence the ongoing development of the mind and/or brain structures." [Wikipedia]

It is perhaps of importance to touch on a brief history of the pre- and perinatal theories. "The relevance of the birth experiences has been recognized since the early days of modern psychology. Although Freud touched on the idea before rejecting it in favor of the Oedipus complex, one of his disciples, Otto Rank, became convinced of the importance of birth trauma in causing anxiety neuroses. It was Rank who developed the process of psychoanalysis based on birth experiences and he authored his seminal work *The Trauma of Birth* on it. Freud's initial agreement and later volte-face of it caused a rift between them which relegated the study of birth trauma to the fringes of psychology. The subject was taken up, again, in 1949 by Nandor Fodor, a patient of Rank's. In addition to birth trauma, Fodor emphasized and wrote much on the significance of *prenatal* trauma.

Developments in the 1950s saw a shift in emphasis toward birth trauma by Donald Winnicott and to the more transpersonal aspects of pre- and perinatal experience by Maarten Lietaert Peerbolte... and brought attention to the relevance of very early gestation and even a focus on the event of conception. These topics saw later elaboration by Frank Lake, as well as Michael C. Irving, R.D. Laing, Graham Farrant, Stanislav Grof and others. The expression at a broad social level of basic perinatal feelings, such as 'suffering fetus' or 'toxic placenta' is part of the narrative in past psychohistory developed by Lloyd deMause. And Pre- and perinatal psychology is at the core of primal therapy and primal integration. And Professor Stephen M. Maret has explored these influences in his book, entitled *The Prenatal Person*." [Wikipedia]. Add to that, Art Janov, a psychotherapist to the stars in Beverly Hills has conducted primal therapy workshops, there, as well.

So, one can readily see the extensive investigation that has already taken place regarding pre- and perinatal influences, to date. And, as the field is further researched and studied, new discoveries regarding this heretofore uncharted period in human development are sure to be further theorized and confirmed. It is perhaps the fastest growing area of behavioral science, today.

The cat and the stove…

Conditioned response plays an integral part in the development of all animals, including humans, and, in fact, is of particular importance in the development of the human psyche.

"Classical conditioning (also Pavlovian or respondent conditioning) is a form of learning in which the *conditioned stimulus* comes to signal the occurrence of a second stimulus, the *unconditioned stimulus* (A stimulus is a factor that causes a response in an organism). The *conditioned response* is then the learned response to the previously neutral stimulus. And the unconditioned stimulus is a biologically significant stimulus such as food or pain that elicits a response from the start, The conditioned stimulus usually produces no particular response, at first, but, after conditioning, it elicits the conditioned response." [Wikipedia]

When I was young, I was given a puppy. In our backyard, we had a barbecue pit and, weather permitting, frequently barbecued on its grill. This required red-hot coals, and when the fire was hot enough, coals would pop out onto the nearby grass. The first day we barbecued after the puppy's arrival, naturally, the little dog wanted to be near the action and unwittingly ran toward the vicinity of the barbecue pit. When his feet stepped on the hot coals, he yelped with excruciating pain and darted away. For the rest of that dog's life, he never went near that barbecue pit, again. If he wanted to go to the back of the yard, he followed the fence perimeter as close as possible until he reached the back. This was a perfect example of classical conditioning, the conditioned stimulus being the hot coals and the conditioned response, brought on by the pain, was to avoid the area, altogether.

Mark Twain said it best: "There was the cat that sat on a hot stove and never sat on a hot stove, again. It never sat on a cold one, either."

Conditioned response serves all animals, well. It protects both humans and non-human animals from experiencing the same pain that caused it in the first place, regardless of whether it is physical or mental. It causes both conditioned and unconditioned (or innate) reflexes.

Current Russian physiologists Nikolai Agajanyan and V. Tsyrkin have postulated that conditioned reflexes, those which are learned, are a function of the cerebral cortex. The cerebral cortex is that which covers the cerebrum and cerebellum of the brain and is divided into the left and right hemispheres. The cerebral cortex plays a key role in memory, attention, perceptual awareness, thought, language and, also, consciousness. The association areas of the cerebral cortex integrate information from different receptors or sensory areas and relate the information to past experiences.

"Neonatal perception is the study of the extent of somatosensory and other perceptual systems during pregnancy. In practical terms, this means the study of fetuses. None of the accepted indicators of perception are found to be present in embryos. Neonatal perception particularly involves the sensing of pain, both physical and mental. However, neonatal scientists are divided on just when a fetus can perceive pain. Other sensory perceptions have been confirmed. Numerous studies have found evidence indicating a fetus's ability to respond to auditory stimuli. Research at Zhejiang University in China indicates that fetuses, at term, can not only hear, but also distinguish their mother's voice from others.

The hypothesis that human fetuses are capable of perceiving pain in the early stages of a pregnancy has not received sufficient evidence to be proven or disproven; the developmental stage of any of the research and its instrumentation is so far insufficient to this task. Some authors, however, argue that fetal pain is possible from the second half of pregnancy.

Researchers from the University of California, San Francisco, noted, in the *Journal of the American Medical Association,* a meta-analysis of data from dozens of medical reports and studies that fetuses are unlikely to feel pain until the third trimester of pregnancy. There is, however, emerging consensus among developmental neurobiologists who have pointed to the establishment of thalamocortical connections (at about 26 weeks) as a critical event with regard to fetal perception of pain. Because pain can involve sensory, emotional and cognitive factors, it may be impossible to determine when painful experiences are perceived, even if it is known when heretofore thalamocortical connections are established." [Wikipedia] Taking into account current scientific assumptions, then, certainly, by the seventh to eighth month of pregnancy, a fetus is well on its way to cognitive perception, the sensing of pain and conditioned response to it.

Based on this evidence, what is taking place in the brain of the mother is likely to be transmitted to the brain of the fetus via stress hormones in the blood through the umbilical cord. This is made possible with the hypothalamic-pituitary-adrenal axis.

"(This) axis is a complex set of direct influences and feedback interactions among the hypothalamus, the pituitary gland (a pea-shaped structure located below the hypothalamus) and the adrenal (also called 'suprarenal') glands (small, conical organs on top of the kidneys). The interactions among these critical organs constitute the HPA axis, representing a major part of the neuroendocrine system that controls reactions to stress and regulates many body processes including digestion, the immune system, mood and emotions, sexuality and energy storage expenditure. This is the common mechanism for interactions among glands, hormones, and parts of the mid-brain that mediate the general adaptation syndrome, as proven..." [Wikipedia]

Therefore, the stress experienced by the mother is likely to be transmitted as a result of the HPA axis via the blood through the umbilical cord to the fetus. And, in the case of stress, paticlarly that which is perceived by the fetus resulting from external stimuli. In other words, if the mother is experiencing emotional pain through stress from problems in, say, a love or close relationship during late pregnancy, this stress will be transmitted to the fetus, and, by extension, perceived to be related to the individual causing that stress, the mother's close relationship or *other*. It is a conditioned stimulus which produces a conditioned (or learned) response.

Falling in love again...

In an overwhelming number of cases, conception is the result of intimacy (or love), although there are situations when it is the result of pure lust and some strictly unwanted encounters... even violent ones as in rape. I will leave immaculate conceptions, however, to those far more spiritual than most of us. Save it to say, that conception is usually the result of a romantic relationship. Let us explore that further.

"During the initial stages of most romantic relationships, there is more often than not an emphasis on emotions – especially ones consisting of love, intimacy, compassion, appreciation and affinity as opposed to physical intimacy, alone.

Anthropologist Helen Fisher, in her book *Why We Love,* uses brain scans to show that love is the product of a chemical reaction within the brain. Norepinephrine and dopamine, among other chemicals, are responsible for excitement and bliss in humans and, to some extent, non-human animals. Fisher concludes that these reactions have a genetic basis, and, therefore, love is a natural drive as powerful as hunger.

In his book *What Women Want, What Men Want,* anthropologist John Townsend takes the genetic basis of love further by identifying how the sexes are different in their predispositions.

Townsend's compilation of various research projects concludes that men are susceptible to youth and beauty whereas women are susceptible to status and security. These differences are part of a natural selection process where males tend to seek healthy women of childbearing age which will mother offspring, whereas women seek men who are willing and able to take care of them and their ospring.

Psychologist Karen Horney, M.D., in her article *The Problem of the Monogamous Ideal,* indicates that the overestimation of love leads to disillusionment, and the desire to possess the partner results in the partner wanting to escape. And the taboos against sex result in non-fulfillment.

Disillusionment plus the desire to escape plus non-fulfillment result in a secret hostility which causes the other partner to feel alienated. Secret hostility in one and secret alienation in the other cause the partners to secretly hate each other. This secret hate often leads one or the other or both to seek love objects outside the marriage or relationship.

Psychologist Harold Bessell, Ph.D, in his book *The Love Test,* reconciles these opposing forces, noted by the above researchers, and shows that there are two factors that determine the quality of a relationship. He postulates that people are drawn together by a force which he calls 'romantic attraction' which is a combination of genetic and cultural factors. This force may be weak or strong and may be felt to different degrees by each of the two love partners. The other factor is 'emotional maturity' which is the degree to which a person is capable of providing good treatment in a love relationship. It can thus be said that an immature person is more likely to overestimate love, become disillusioned and have an affair, whereas a mature person is more likely to see the relationship in realistic terms and act constructively to work out problems.

Romantic love, in the abstract sense of the term, is traditionally referred to as involving a mix of emotional and sexual desire for another as a person. However, Lisa M. Diamond, a University of Utah psychology professor, proposes that sexual desire and romantic love are functionally independent and that romantic love is not intrinsically oriented to same-gender or other-gender partners. She states that the links between love and desire are bidirectional, that is, qualifying as being opposed to being unilateral. Furthermore, Diamond does not state that one's sex has priority over the other, male or female, in romantic love and her theory suggests it is as possible for someone who is homosexual to fall in love with someone of the other gender as for someone who is heterosexual to fall in love with someone of the same gender, as well. In her review of this topic, Diamond emphasizes that what is true for men may not be true for women. According to Diamond, in most men, sexual orientation is fixed and most likely innate, but in many women sexual orientation may vary from Kinsey 0 to Kinsey 7 and back, again.

Martie Haselton, a psychologist on staff at UCLA, considers romantic love a 'commitment device' or mechanism that encourages two humans to form a lasting bond. She has explored the evolutionary rationale that has shaped modern romantic love, in general, and she has concluded that long-lasting relationships are helpful to ensure that children reach reproductive age and that they are fed and cared for by two parents. Haselton and her colleagues have found evidence, in their controlled experiments, that prove or suggest love's adaptation. The beginning part of the experiment consists of having people think about how much they love someone then suppress thoughts of other attractive people. In the second part of the experiment the same people are asked to think about how much they sexually desire those same partners and then try to suppress thoughts about others. The results showed that love is more efficient in pushing out those rivals than sex is.

Research by the University of Pavia, in Italy, suggests that romantic love lasts for about one year and then it is usually replaced by a more stable form of love referred to as companionate love. In companionate love, changes occur from the early stage of love to when the relationship becomes more established and the person's romantic feelings seem to end. However, research done at Stony Brook University in New York suggests that some couples keep romantic feelings alive for much longer." [Wikipedia]

Whatever the case, it is a natural human reaction and is evident in non-human animals, as well. "Falling in love again, never wanted to… what am I to do; I can't help it…" so perfectly expressed in the song written by German-born Friedrich Hollaender and popularized in Western culture by Marlene Dietrich.

One thing is for certain, romantic love is present in a variety of levels from strong to weak or even non-existent, with a stress factor going from a point of low to high, respectively. Conception can occur at any level, of course, including the virtually non-existent; that is, no romantic feelings for the partner, at all. How then might this be perceived by the fetus which is influenced by stress hormones coming from the mother due to her current state of mind?

A fetus is, for all intents and purposes, in a *state of bliss*. The warmth of the womb and ample nourishment create an ideal environment in which the fetus luxuriates from embryo to birth… a state of contentment until something invades its realm, like *stress*. "Oh, just when I was beginning to enjoy this heavenly state, now this!"

If the father or a male partner has remained around by at least the seventh or eighth month of pregnancy, the fetus soon begins to perceive the presence of another (cognitive perception). If that being's presence is romantically strong and likewise wanted by the mother, the mother's emotional stress level is apt to be low. Very little stress is transmitted to the fetus (only that from normal causes) and the state of bliss is hardly disturbed. As a result, the fetus does not perceive this other's presence to be a threat to its blissfulness. And, if heterosexual sex is continuing as a part of that relationship, the fetus perceives this activity (the intrusion of a penis into its near-immediate realm) and accepts it as non-threatening, as well. It becomes for the fetus, over time, accustomed and becomes acceptable. In fact, the fetus begins to associate this physical intrusion with the bliss state, and the conditioned response is positive and even reinforced. The conditioned stimulus, the *other*, together with the penis and the mother having an absence of or minimal stress, produces a combined conditioned stimulus and positive response. The fetus, therefore, accepts this *other*, not only as welcomed, but gratifying.

Although this continually stress-free existence contributes to the ideal environment for the fetus, it does not always exist. Remember that romantic relationships come in levels of feelings strong to weak and even non-existent. The more positive or stronger the bond between the mother and the *other*, her love partner, the more blissfully-gratifying it is for the fetus. And this degree of attachment is going to have a direct effect on the fetus's later development during postnatal maturing. I say the *other* which may or may not be a father or other male partner because, at this pre-birth state, the fetus is not yet aware of gender opposites. It is still in the unconscious state, the id, as Freud called it. It only perceives the presence of the *other* and that other often possesses a penis.

If the relationship, however, between the mother and the other is strained during the last two months of pregnancy, then the stress level of the mother rises, as a result. Regardless of whether the strain caused in the relationship comes as a factor of the emotional distancing by the mother or emotional distancing by the *other*, either way, stress is apt to be created in the mother's brain and, through the umbilical cord, is transmitted by hormones directly to the fetus. *Stress is the key factor.* It is stress that upsets the state of bliss. In other words, the fetus perceives stress as a negative conditioned stimulus coming in conjunction with the presence of the *other* and a negative conditioned response to this is the result. How often that stress continues throughout those last two months is the degree to which that conditioned response is reinforced.

If the *other* drifts away, emotionally and/or physically (meaning no physical connection or penile intrusion occurs) during this two-month period, then the conditioned response will be that 'nothing is better than something negative.' In this case, the fetus perceives nothing to be a continuation of the bliss. We can define this occurrence, then, as a 'perception of non-existence' of *other*, father or other male or female partner. That is not to say, however, that this absence, in itself, is non-stressful for the mother.

A little Greek history, please…

"The term 'Oedipus complex' denotes emotions and ideas that the fetus keeps in the unconscious, via dynamic repression, and concentrates upon the child wanting to sexually possess the parent of the opposite sex. Freud, who coined the term, believed it to be a desire of the parent by both sexes. 'Oedipus' refers to the 5th century B.C. mythological Greek character named Oedipus who unwittingly kills his father, Laius and marries his mother, Jocasta.

In classical Freudian psychoanalytic theory, a child's eventual identification with the same-sex parent is the successful resolution of the Oedipal complex… so, key psychological experiences that are necessary for one's development of a mature sexual role and identity. Freud further proposed that boys and girls experience the complex differently, boys in a form of castration anxiety and girls in the form of penis envy. And that unsuccessful resolution of the complex thereby leads to neurosis, pedophilia and homosexuality. In adult life, this can also lead to a choice of a sexual partner who resembles one's parent.

The Oedipus complex, according to Freud, occurs during the phallic stage of psychosexual development (age 3-6 years), when also occurs the formation of the libido and the ego; yet it might manifest itself at an earlier age.

In the phallic stage, according to Freud, a son's decisive psychosexual experience is his son-father competition for possession of mother. And, in females, competition for possession of father. It is during this third stage of the psychosexual development that a child's genitalia are his or her primary erogenous zone. Thus, when children become aware of their bodies, the bodies of other children and the bodies of their parents, as well, they gratify physical curiosity by undressing and exploring themselves, each other and their genitals... learning the anatomic differences between male and female and the gender differences between boy and girl.

In countering Freud's proposal, that the psychosexual development of boys and girls is equal, that each initially experiences sexual desire for mother and aggression toward father, Jung proposed that girls experience a desire for father and aggression toward mother, instead, via the Electra complex derived, alternatively, from the 5th-century B.C. Greek mythological character Electra who plotted matricidal revenge with her brother Orestes against Clytemnestra, their mother, along with Aegisthus, their step-father, for their murder of Agamemnon, her real father, (by Sophocles)." [Wikipedia]

Contemporary psychoanalysts accept as fact the universality of the Oedipus complex, to different degrees, that the child enters an Oedipal phase of acute awareness of a complicated triangle involving mother, father and child, and that both positive and negative Oedipal themes are observable in development. To Freud, the unresolved complex determines sexual preferences and orientation. I postulate that, what it does is allow preferences and orientation to manifest, later, in the adult sexual stage. The determining of preferences and orientation, I, contrarily, believe, actually occurs much earlier. The determining is made prenatally while the fetus is in the womb. We will come to understand this more, later.

Can we ever go home...?

In 1919, Freud wrote one of his most far-reaching articles on sexuality which he entitled *The Uncanny*... that which is mysterious and totally unknown to him, even foreboding. In the article, he lays out the most 'uncanny' experience he encounters, the 'female womb,' that blissful, heavenly comfort-zone where man first began life and the experience of which is something he never completely comprehends, except in the aesthetic.

In this treatise, Freud associates man's 'wants and desires to be in love' with a longing to return home... in other words, to return to the womb for which man strives on an unconscious level to replicate through sexual intercourse. But the effort is, of course, ever futile for, with orgasm, he shrivels away from his quest and is left with a continual longing and unable to remain there to be nurtured, forever. And it is, as Freud states, true for both male and female. "You can never go home," wrote the author Thomas Wolf... and Freud laid this out, precisely, in a psychosexual context.

Nature provides this quest to return, this drive to go symbolically home, as a basic instinct. It is what guides every living creature to return to its beginning to procreate... renew... to pass on. Geese fly thousands of miles to where they were first hatched to mate and rejuvenate. Salmon swim hundreds of miles upstream, maneuvering nearly-impossible rapids, to the shallow pools where they began life to spawn. Mammals return to the womb in an act of sexual intercourse for exactly the same reason... steered by instinct. It is the one drive that man is never able to fulfill, but nature accomplishes her aim, nonetheless... procreation... and life goes on.

In humans, this unfulfilled longing manifests itself, later, in a tumultuous complex in a child's life. Freud described the source of this complex in his *Introductory Lectures*... He began with, "You all know the Greek legend of King Oedipus who was destined, by fate, to kill his father and take his mother as his wife, who did everything possible to escape the oracle's decree then punishes himself by blinding when he learns he has, unwittingly committed both these crimes..."

According to Freud, Sophocles' drama illustrates a formative stage in the life of every individual's psychosexual development when the child transfers his or her focus from the breast, the oral phase, to the mother, herself. At this time, the child desires the mother and resents, even secretly desires the murder of the father. For the male, the Oedipus complex is closely related to what Freud called the castration complex. Such primal desires are naturally repressed, but even among the mentally sane, they arise in dreams or in literature. Among those individuals who are never able to progress, properly, to what Freud calls the genital phase, the unresolved complex can lead to the playing out of a psychodrama in displaced, abnormal and/or exaggerated ways. Herein lies, I believe, the source of rape.

So, this desire to return home, this *womb obsession*, as I term it, and as nature has skillfully planned, carries out the necessity for the continuation of life... and of the species. However, man's advanced intelligence has led to states of anxiety and confusion, as a result of it.

In his book entitled *Man's Presumptuous Brain*, an evolutionary interpretation of psychosomatic disease by Albert T.W. Simeons, the author states that "Civilization is the method of perfecting artificial means of escape from biological dangers and imperatives through the brain cortex, the newest part of it. Civilization has gone so far that it virtually prevents urban man from reacting to danger in a way normal for the individual, evolutionarily speaking. Modern man is so impressed and overwhelmed with the achievements of his neo-cortex that he forgets his body still functions on a level normal before the dawn of time. Modern man has evolved so far away from the diencephalic reactions of his body that they no longer make sense to his neo-cortex. And he misrepresents these responses as disease. We thereby lay down the pattern, here, that eventually leads us into psychosomatic suffering."

This may slightly overstate the issues but, suffice it to say, the developing fetus, still at a state of primal perception and response, or id, views its world of tranquility as bliss, and any disturbances to that realm are taken to be negative, even violent. The following reaction is predictable... avoid the disturbance, at all cost.

We have already seen where stress, caused by the mother's state of mind, is transferred by stress hormones through the umbilical cord to the womb, rendering the placenta toxic and upsetting the earlier tranquil existence, within. This may not be that disturbing, initially, but, as we have seen, frequent and/or repeated occurrences reinforce the conditioned response. Eventually, the fetus must make a decision... a fundamental choice for its survival... a 'fight or flight' moment to this unwanted state. "Stress, stress, stress... will somebody give me a break!"

"The fetus develops inside the womb uterus with the help of a fetal life-support system composed of the placenta, the umbilical cord, and the amniotic sac...

The placenta has been described as a pancake-shaped organ that attaches to the inside of the uterus and is connected to the fetus by the umbilical cord. The placenta produces all the pregnancy-sufficient hormones, that it needs, including chorionic gonadotropin (hCG), estrogen, and progesterone.

And the placenta is responsible for working as a trading post between the mother's and the fetus's blood supply. Small blood vessels carry fetal blood through the placenta which carries maternal blood, as well. Nutrients and oxygen from the mother's blood are transferred to the fetal blood while waste products are transferred from the fetal blood to the maternal blood without the two blood supplies mixing.

The umbilical cord is the lifeline that attaches the placenta to the fetus. It is made up of three blood vessels: two smaller arteries which carry blood to the placenta and a larger one which returns blood to the fetus. It can grow to be 60 cm long, allowing the fetus enough cord to safely move around without causing damage to it or to the placenta.

The amniotic sac is filled with the amniotic fluid. This sac is the fetus's home and protection from outside knocks, bumps, and other external pressures. The amniotic sac allows the fetus ample room to swim and move around which helps build muscle tone. To keep the fetus cozy, the amniotic sac and fluid maintain a slightly higher temperature than the mother's body, usually about 99.7 F." [American Pregnancy Association]

We have already seen where stress hormones from the mother pass into the fetus's environment through the umbilical cord. Just what effects might occur as a result of this process? "A study done at the Hirosaki University Graduate School of Health Sciences in Japan examined the correlations between perceived stress-associated hormone/oxidative stress markers in umbilical cord blood and the physical condition of the mother and neonate, and found that CB stress-associated hormone/oxidative stress markers strongly reflect maternal and neo-natal condition, at the time of delivery.

Depression during pregnancy has many health implications, including a new effect seen in babies born to depressed mothers. Research from the University of Michigan School of Medicine has found that these babies have higher levels of stress hormones, decreased muscle tone and additional neurological and behavioral differences. In the study, higher levels of depression in mothers, during pregnancy, was associated with higher levels of stress hormones in their offspring, at birth, as well as with other neurological and behavioral differences.

The analysis, which appeared in *Infant Behavior and Development,* examined various links between maternal depression and the development of an infants' own neuroendocrine system that controls the body's stress response as well as moods and emotions. The longer-term question for researchers is the degree to which the hormonal environment in the uterus may act as a catalyst for processes that, in turn, alter infant gene expression, neuroendocrine development and brain circuitry, potentially setting the stage for increased risk for later behavioral and psychological disorders.

Along with them tracking the mothers' depressive symptoms throughout gestation, the researchers took samples of umbilical cord blood, right after birth. They found elevated levels of adrenocorticotropic hormone (ACTH) in babies born to mothers with depression. This hormone tells the adrenal gland to produce the stress hormone cortisol.

The impact of mothers' depression on fetuses and newborns has generated a considerable amount of research, in recent years. Previous studies have shown that babies born to mothers with severe depression may be more likely to birth prematurely or underweight, have diminished hand-to-mouth coordination and be less cuddly." [National Institute of Health]

So, we have seen how conditioned response operates on a subconscious level in humans as well as non-human animals. Remember the cat on the hot stove... and the cold one, too? We have likewise seen that conception often occurs as a result of a love relationship lasting right up until birth... but lasting that long is not always the case. Relationships go sour and may even terminate before that critical period in fetal development. And we have also seen where the psychosexual development of the infant, following birth, goes through a series of psychosexual development stages, defined as the Oedipus (or Electra) complex which presents the child (age 3 to 6) with a triangular dilemma of mother, father and child, and the trauma which that causes the child that must later be resolved in order to permit a healthy psychosexual development. Now, we are beginning to see how stress, possibly caused by a relationship going sour or terminating, altogether, in late pregnancy, can have a direct and critical effect on the fetus. But just what effects will our vulnerable fetus suffer?

Nature, as we have also learned, has provided all living creatures with instincts... instincts that are crucial for not only survival but also for carrying on the species. And these instincts operate at the subconscious level (the id), a primordial condition that began as far back as when all living creatures were simply single cells. Through millions of years of development, each species continued with these instincts and creating new ones as the being metamorphized into a fully-functioning animal. But the fetus, unlike the born offspring, is still in the primordial stage, operating on a subconscious level, controlled by these very basic instincts.

We have also seen where the realm of the fetus, the womb, is a blissful state, warm, nurturing and without little or no negative occurrences. Even with the presence of the *other*, that being outside the womb, outside the mother, doesn't appear to upset this state of bliss, to any degree. Even when there is heterosexual sex and intrusion of a penis into the fetus's immediate environment, this, too, is readily accepted as non-threatening. The presence of this strange member, which frequently makes its way up the vaginal canal toward the very door to the womb, appears to the fetus to be completely benign.

But what happens when this is not the case... when stress and often high levels of it is perceived by the fetus? The fetus soon realizes the state of bliss has been altered... is less desirable... and even uncomfortable. At its subconscious level, its conditioned response instinct kicks in. The stress is perceived as a change for the worse, and it is, therefore, negative. This stress is the first conditioned stimulus the fetus ever encounters. And, it is nature providing it with its first conditioned response... avoid this at all cost. And, if this stress continues, repeatedly, throughout the last formative two months in the womb, during cognitive perception, the conditioned response is, as we have seen, reinforced.

By the time birth occurs, the fetus has already had this negative response locked into its subconscious to protect it from wearily of ever occurring, again. But it will have to be confronted, later, one day... when the child reaches sexual maturity. Its unconscious memories will surface and its sexual preferences and orientation will be revealed, as a result of it.

For the time being, though, the newborn is rapidly developing its new conscious mind, its ego. And, as the child encounters the cultural world around it, the super-ego, as well. "Oh, the trials and tribulations of my growing up years." Thanks to nature, though, this developing youth will not have to deal with advanced psychosexual complications until puberty, when further trauma occurs.

Philadelphia will do...

There is still a critical piece of our sexual preferences and orientation puzzle to connect.

"Adolescent sexuality refers to sexual feelings, behavior and development in adolescents and is a stage of human sexuality. It is often a vital aspect of teenagers' lives. The sexual behavior of adolescents is, in most cases, influenced by their culture's norms and mores, their sexual orientation acceptances and issues of social control such as age of consent laws.

In humans, mature sexual desire usually begins to appear with the onset of puberty. And sexual expression can take the form of masturbation or sex with an equally-aged partner. And sexual interests among adolescents, as among adults, can vary, greatly. Sexual activity, in general, is associated with various risks, however, including unwanted pregnancy (in mid to late adolescent years) and sexually transmitted diseases, including HIV and AIDS. The risks are elevated for adolescents because their brains are not neutrally mature; several brain regions in the frontal lobe of the cerebral cortex. The hypothalamus, important for self-control, delayed gratification and risk analysis and appreciation, are not fully mature. The creases in the brain, also, continue to become more complex until the late teens and the brain is not fully mature until age twenty-five. Partially because of this, young adolescents are generally less equipped than adults to make wise and sound decisions in anticipating consequences in their sexual behavior. Nonetheless, brain-imaging and behavioral correlation studies in teens have been criticized for not being causative and possibly reaffirming cultural biases." [Wikipedia]

During this adolescent period, the individual likewise rediscovers self. The first discovery, as Freud pointed out, is the genital stage when the child first discovers his or her body sexually and explores self as well as the bodies of others. But, at this stage, the brain's development is still infantile and the child is not fully capable of his or her eroticization. Only with puberty will he or she achieve this. What Freud called simply the genital stage, during this Oedipal period, I introduce a second, delayed self-discovery period, post puberty which I call the *latent* genital stage. Here, the soon-to-be adult is now fascinated with his or her developing genitalia, all over again, along with the genitalia of same sex others (called gender identification) now combined with eroticization and full maturity of the genitals. It is a natural same-sex stage in which teenagers begin to experiment with same sex partners… sort of a learning stage, preparing them for adult heterosexual behavior, in the future. This stage can be early, on time or delayed. In any case, it is a normal process in human psychosexual experimentation.

But, as the individual advances toward adult sexuality, at the rate of a speeding freight train, little does he or she realize the head-on collision waiting down the line. For, in the subconscious mind, that still primeval id is determined to loosen a few ties and cause havoc with the individual's conscious reality... and that waiting collision is with the opposite sex. During the prenatal stage, the fetus was unaware of gender opposites. As we have seen, it simply perceived the *other* outside the mother. Following birth and during the first genital stage, the child quickly begins to distinguish between male and female as the child's ego matures. During the second or latent genital stage, the adolescent, fascinated by same sex genitalia, has still not become fully eroticized by the opposite sex (gender-opposite identification). When this begins to occur, all hell breaks loose. Those shifting ties begin to upset the safe haven of the adolescent mind and a full train wreck is inevitable. "Oh my... how do I deal with this?"

As Freud and Jung pointed out, during the Oedipus stage the male child desires the mother and resents the father and the female the father resenting the mother. This latent desire has been suppressed for the child knows, all too well, this is unacceptable... even forbidden. The child may have fantasized about it, either consciously or subconsciously, but it remains an unreachable desire.

Here, nature comes to the rescue, once again. For all humans and non-humans alike, the ability to substitute one's wishes, one's fulfillments, at this stage, can and usually is accomplished by, what I have defined, earlier, as *objective representation*... not to be confused with Freud's transference, in psychoanalysis, which occurs between patient and therapist. This representation utilizes a surrogate... and, in this the case of this, a surrogate parent; in specific, a *substitute* womb.

All animals are capable of opting for objective representation. Dogs will often play with stuffed animals, for in the subconscious mind, there is no distinction between the living and non-living... only in comfort... in feelings. Cats play with movable objects as if that were living prey. And children become intimately attached to dolls and teddy bears, even talking to them as if these were living beings, like themselves. Some children even create imaginary friends with whom they converse, freely. In other words, these representations or substitutes supply a comfort need. As W.C. Fields once remarked, "I'd like to see Paris before I die... Philadelphia will do." These substitutes for the individual not only do well... they do quite well.

So, the adolescent who suppressed those early desires to possess a parent, sexually, is now fully eroticized in his or her developmental and must come up with a viable solution... a substitute womb which will satisfy those desires now that the adolescent is ready to move on them. In the case of the male, that will become a female other as a substitute for the mother... a substitute womb. And, in the case of the female, her substitute father. This is the natural progression toward heterosexuality for those whose prenatal experiences were blissful... in other words, where there was virtually little or no stress in the womb and where the conditioned response was positive or infrequent. With his own penis fully mature, now, and preoccupying virtually his every thought feeling, the male will pursue his journey home through objective representation, another female. And, in the case of the female, using her own objective representation, another substitute father's penis in another male. In her own way, then, the female will pursue, subconsciously, her own journey home.

But what happens in the case of adolescents whose prenatal womb experience was stressful? Their conditioned stimulus was, on the contrary, negative, and the conditioned response was to avoid that experience, at all cost. The toxic placenta (highly stressed) eventually rendered the womb undesirable. He or she, though, instinctively longs to go home. So, how will this be accomplished if the pursuit is through the normal act of intercourse?

In this case, the individual will, subconsciously, reject the idea of the mother's toxic womb as undesirable, at all cost, and will opt for the *other* in the prenatal trilogy (mother, other and child). This *other* will serve in the subconscious as the most acceptable alternative, of the three. And that surrogate will possess a substitute like-womb, the anus. Through objective representation, the male, free now of womb toxicity, attempts to go home.

And, in the case of the female, she will opt for the other, as well, free of the mother's toxicity, for the most appropriate *other* (Remember, to the fetus, it is still unaware of gender opposites, only aware of the *other*), free of toxicity, also choosing to avoid penis invading her realm. In either case, the goal is the same... to return home... to return to the state of bliss. This alternative choice, in both sexes, is what is referred to as homosexual orientation and is fixed for life as the subconscious mind is not adaptable in the way the conscious is, through its continual cognitive learning process.

There is something more which needs further clarification. There are no opposites within the subconscious for homosexual or heterosexual states. It is a continuum of degrees. On a scale of one to ten, there is no one or ten. It is a degree from two to nine, and homosexual being 'two to five' and heterosexual 'six to nine.' And this scale of positioning is most definitely influenced by the degree to which stress was experienced in the womb. It is thought to be more pronounced in the male, however, who is conceived both male and female, genetically (X+Y chromosomes) whereas the female is conceived only female (X+X).

We have seen that romantic love comes in degrees. And, with problems in the mother's romantic relationship, stress rises... and, with serious amounts of it, becomes extreme. But, in a mid-range stress level, this most likely causes an ambiguity in the conditioned response. The fetus is teetering on one way or the other: "Is stress really all that bad or is it tolerable?" In the subconscious of the fetus, psychosexual orientation may be in a range of five to six and fluctuate back and forth at five to six. This orientation is referred to as bisexuality. In popular terminology, it is referred to as 'fence sitting.'

One more area that needs explanation... that of preferences. This is a term that is used, frequently, to convey a variety of sexual likes or dislikes... fetishes, really. But, in this case, I am referring to power positioning. We saw in the womb, earlier, the fetus perceiving the intrusion of a penis in the vaginal canal, right up to the door of the womb, as if banging on it. We also saw that, in the case of a virtually free or low stress environment, the *other* was perceived as non-threatening. Still operating on a subconscious level, the fetus connected that intrusion as accepting, but connected it, nonetheless. It became an integral part of the fetus's cognitive awareness.

In this regard, when the fetus developed into the mature stage of psychosexuality, the penile memory was still there in the id. Whenever it was introduced, during the last two months of pregnancy, it was a powerfully aggressive act which the fetus married in its subconsciousness with its continual stress-free bliss. No or little stress hormones had entered its realm. At the mature stage in pregnancy, that penile intrusion is made aware. In the case of the male, he will become dominant in sex which will be his norm. If no aggressive intrusion was perceived, he will not have associated aggressiveness with the act and will become passive... and that will be his norm. In other words, his role will be submissive in regard to his female partners or he will simply have little interest in sex, in general.

In the case of the female, where penile intrusion occurred, she, too, will have associated aggressiveness with the sex act and will become dominant in her love relationships. For those females where there was no aggressive intrusion or it was infrequent, she will not have associated aggressiveness with the act and her norm will be submissive.

So, that leaves us to compare fetuses who experienced high stress and/or toxic levels in the womb and where the conditioned response was negative… and where they have chosen an alternative route to go home. In the case of the male, where penile intrusion was frequent, he will have associated aggressiveness with the act and will become aggressive in his homosexual activity… in other words, dominant. He will take the penile approach in his sexual activity. In popular terminology, this is referred to as *top*.

And, for those homosexual males whose penile experience had no (or little) intrusion, he did not associate aggressiveness with the act and will be passive or submissive. And, like the heterosexually-oriented passive female, who chose the other without aggression, he will adopt his anal canal as the most appropriate *other*, and seek penile intrusion into his own womb-like orifice. And in popular terminology, it is called *bottom*.

In females where penile intrusion was frequent, she will have associated this aggressiveness with the act, as well, and will take the dominant role with her female partners. And where no penile intrusion (or little) was perceived, she will remain passive, taking a submissive role.

And when you reach the end...

...said the king to Alice... "then stop!" How apropos.

What I have laid down for you is simply theory. *"Everything we hear is an opinion, not a fact. Everything we see is a perspective, not the truth."* Those were the wise words of the Roman emperor Marcus Aurelius. For, with every theory, there is dissent. Researchers and scientists strive to prove or disprove each one but, in the end, we are ever skeptical. Some theories *are* proven, without a doube and that is certainly the aim. The world is round, not flat. That has been proven. Einstein gave us his theory of relativity but scientists are still trying to prove it. In the case of human behavior, there never appears to be a resolution, of certainty. All, in this regard, remains a mystery. Freud, of course, gave us many theories, including his haunting Oedipus complex. To this day, behavioral scientists are still debating it. In the field, one is apt to refer to him- or herself as *Freudian* or *not*, depending on their personal affirmations regarding Freud, himself. We may never know the truth, only a perspective.

Stress hormones inhibiting testosterone production in fetuses have been studied by a host of researchers, including Dr. Gunter Dorner of Germany. Dorner found a significant-increase in male homosexual births in East Germany between 1941 to '47, during widespread Allied bombings and stressed pregnancies. His team further interviewed one hundred bi- or homosexual men about the occurrence of maternal stress relayed to them during their prenatal period and, likewise, found similar results. This led Dorner to conclude, in addition to his earlier research on testosterone-deprived rats, that stress-induced low testosterone levels most likely had a role in bi- and homosexuality. Other studies on lack of testosterone during pregnancy have not yielded the same results.

Dr. Jacques Balthazart, at the University of Liège in Belgium, one of the world's foremost and prolific behavioral neuroendocrinologists, reports a host of research he has conducted on hormonal deficiencies during pregnancies. He has recently published a book on the topic, entitled *The Biology of Homosexuality*. His conclusions favor the biological deficiency angle, and he has also contributed to my research finding my assessments compelling. I stay clear of hormonal deficiencies, however, which may or may not affect physical masculinization, only. Instead, I focus on the cause of bi- and homosexuality as a direct result of stress-induced womb toxicity during the last two months of pregnancy and the negative conditioned response it induces.

So, the theory of womb obsession is merely a perspective. It is not for me to claim it as fact, only as theory. It is the role of behavioral scientists to put it to the test. Research will be needed, of course. But, even with that, we may never be able to proclaim it unconditionally true.

One thing continues to weighs heavily on my mind; overwhelmingly, those who have strong attractions for their same sex partners claim they were born that way. For them, it is innate. Freud believed it was the result of the unresolved Oedipus complex. Others have claimed early life experiences, dominant mothers, submissive fathers... the list goes on. I believe it is innate (meaning at birth), as well... and those preferences and orientation have been formed in the subconscious during the stressful prenatal stage. Only further research can shed additional light on the subject, and I welcome that.

Paperback copies available from Amazon:

https://www.amazon.com/Womb-Obsession-Theory-Preferences-Orientation/dp/B086PLXXXH/ref=sr_1_17?crid=LCUERRKYMWZ5&keywords=claude+brickell&qid=1656557154&sprefix=%2Caps%2C103&sr=8-17